FAMOUS NAVY FIGHTER PLANES

GEORGE SULLIVAN

DODD, MEAD & COMPANY
New York

PICTURE CREDITS

The Boeing Company, 11; LTV Aerospace and Defense Company, 6, 15; McDonnell-Douglas, 59 (bottom); The Northrop Corporation, 61; National Air and Space Museum, Smithsonian Institution, 12, 13, 21, 26, 27, 42, 43, 47; George Sullivan, 58, 63. All other photographs are Official U.S. Navy Photographs.

Copyright © 1986 by George Sullivan
All rights reserved
No part of this book may be reproduced in any form without permission in writing from the publisher
Distributed in Canada by
McClelland and Stewart Limited, Toronto
Manufactured in the United States of America

1 2 3 4 5 6 7 8 9 10

Library of Congress Cataloging-in-Publication Data

Sullivan, George, 1927–
   Famous navy fighter planes.

   Summary: Traces the development of Navy fighter planes by examining specific types from World War I to recent times.
   1. Fighter planes—United States—History—Juvenile literature. 2. United States. Navy—Aviation—History—Juvenile literature. 3. Aeronautics, Military—United States—History—Juvenile literature.
   [1. Fighter planes—History. 2. Airplanes—History. 3. United States. Navy—Aviation—History] I. Title.
UG1242.F5S5919   1986   358.4'3   86-6202
ISBN 0-396-08769-8

INTRODUCTION

Military flying machines dating to the early 1900s were not designated as either fighters or bombers. About all that planes of that period were supposed to do was fly. No one thought to put them into categories.

World War I, which began in Europe in 1914, changed all that. The warring nations, the British, in particular, developed many different types of aviation equipment and operating techniques. By war's end in 1918, navy aircraft could be divided into these classifications:

- Fighters, to destroy enemy aircraft in the air and protect bombers.
- Bombers, to carry and drop bombs.
- Torpedo attack, to carry and launch torpedoes.
- Reconnaissance, to search for and gather useful military information.
- Trainers, to train pilots and crew members.

World War I triggered not only great change but tremendous growth. In 1917, the year that the United States became an active participant in the war, the Naval flying corps had only 49 pilots, 54 planes, and one air base. None of the aircraft was fit for combat duty.

By the time the armistice ending the war was signed in November, 1918, the strength of the Navy's air arm had expanded to 6,716 officers and 30,693 enlisted men. They operated some 2,107 aircraft, or "aeroplanes," as they were called in those days.

In the postwar years, while the size of the Naval flying corps was reduced, it gained in importance. This was the period in which aircraft came to be looked upon as a major weapon. Six aircraft carriers were in operation by the end of the 1930s and more were being built.

The Navy operated an air arm of enormous size during World War II. At its peak, on July 1, 1945, it consisted of 40,912 aircraft.

Naval aviation put forth another major effort during the Korean War, from 1950 through 1953, flying tens of thousands of missions from thirteen different aircraft carriers.

The Navy swung into action again during the Vietnam War. Attacks from carriers stationed off the coast of Vietnam were a daily routine by 1965 and continued to the war's end.

This book and its companion volume, *Famous Navy Attack Planes*, trace the development of the Navy's flying machines. In not much more than three-quarters of a century, they have progressed from fragile and lumbering craft to sleek, sophisticated jets, which include some of the fastest and most powerful aircraft in the world.

The author is grateful to many individuals who helped him in the preparation of this book. Special thanks are due Lt. J. A. Kendrick, Anna C. Urband, Bob Carlisle, David Kronberger, Office of Information, U.S. Navy; Roy Grossnick, Gwendolyn Rich, John Elliott, Aviation History Office, Office of the Chief of Naval Operations; Phil Edwards, Norman Richards, National Air and Space Museum; Capt. Leighton W. Smith, Jr., USS *America*; Grover Walker, Naval Aviation Museum, Pensacola, Florida; Barbara Weiner, USS *Intrepid* Sea-Air-Space Museum; Jerry Borenstein, Naval Aviation Commandery; Ira Chart, Northrop; R. F. Foster, McDonnell; Marilyn A. Phipps, Boeing; Bob Harwood, Grumman; Francesca Kurti, TLC Custom Labs; and Rear Adm. (Ret.) John J. Schieffelin.

CONTENTS

Introduction	3	Vought F4U Corsair	36
Vought VE-7	6	Grumman F7F Tigercat	40
Naval Aircraft Factory TS-1	8	Grumman F8F Bearcat	42
Boeing FB	10	Ryan FR Fireball	44
Curtiss F6C Hawk	12	McDonnell FH Phantom	46
Vought FU-1	14	Douglas F3D Skyknight	48
Boeing F3B	16	Grumman F9F Panther	50
Boeing F4B	18	LTV F-8 Crusader	52
Curtiss F9C Sparrowhawk	20	Grumman F11F Tiger	54
Grumman FF	22	McDonnell-Douglas F-4 Phantom II	56
Grumman F2F, F3F	24	Northrop F-5E Tiger II	60
Brewster F2A Buffalo	26	Grumman F-14 Tomcat	62
Grumman F4F Wildcat	28	Navy Aircraft on Display	64
Grumman F6F Hellcat	32		

VOUGHT VE-7

Aviation began in America with the breakthrough achieved by the Wright Brothers in 1903. In the years that followed, other nations of the world pursued the idea of the flying machine with greater zeal than did Americans. As a result, by 1914 and the outbreak of World War I, the United States had fallen far behind Britain, France, Germany, Italy, and Russia in aircraft development.

The thousands of American pilots who were sent to Europe during the war had to fly British and French planes, which were largely unreliable. The United States did not produce a single combat-ready aircraft during World War I.

In the summer of 1918, a few months before the

Besides serving as a fighter, the VE-7 was a trainer (in the version shown here) and an observation plane.

war was to end, the Aircraft Production Board urged American industry to develop original aircraft designs. One result was the VE-7, a product of the then new Lewis & Vought Corporation of Long Island City, New York. The firm was later named the Vought Corporation after its founder, Chance M. Vought, a noted airplane designer of the time.

The first VE-7 was completed in the spring of 1920. On May 27 of that year, the aircraft was flown from Mitchel Field on Long Island, New York, to the Anacostia Naval Air Station outside Washington, D.C.

The VE-7 was intended originally to be used only for training purposes. But it performed so well as a two-seat trainer that several other versions of the aircraft were produced. It served as an observation plane and, when fitted with a central float, became a seaplane, able to land on and take off from water.

In one other model, the VE-7S, the front cockpit was covered and a pair of forward-firing Vickers .303-caliber or Browning .30-caliber machine guns were installed. This was the Navy's first American-built fighter. The VE-7s remained as frontline equipment until 1926.

The VE-7 is noteworthy for at least one other reason. On March 20, 1922, the Navy's first aircraft carrier, the USS *Langley*, was commissioned. On October 17 of that year, Lt. Cmdr. V. C. Griffiths became the first pilot to take off from the *Langley*. The plane Griffiths flew was a VE-7.

Other Data (Model: VE-7)
Wingspan: 34 ft., 1 in.
Length: 24 ft., 5 in.
Power Plant: One 180-hp Wright Whirlwind
Loaded Weight: 2,100 lb.
Maximum Speed: 117 mph at sea level

Like many early aircraft, the VE-7 was fitted with pontoons which enabled the aircraft to land on and take off from the water.

NAVAL AIRCRAFT FACTORY TS-1

The U.S. Navy got its first aircraft carrier in 1921. It was not a ship built from the ground up, but a conversion. A coal-carrying vessel—a collier—was turned into an aircraft carrier by constructing airplane hangers and gasoline storage spaces where there had once been enormous coal bins. It was topped with a 542-foot wooden flight deck. The vessel was named the USS *Langley*, in honor of Samuel P. Langley, an aviation pioneer whose model "aerodrome" machines were flying successfully as early as 1896.

At the time the ship was being converted, the officials at the Bureau of Aeronautics in Washington realized that the Navy had no flying machine specifically designed to operate from a carrier flight deck. Engineers at the Naval Aircraft Factory in Philadelphia were called upon to design a plane to fill the need.

The result was the TS-1, a single-seat biplane that was designed to operate either with wheels or

The TS-1 was the first aircraft of the U.S. Navy designed and built as a fighter plane.

The seaplane version of the TS-1 operated with Navy destroyers, cruisers, and battleships.

as a twin-float seaplane. Fuel was carried in a hollow section in the center of a lower wing.

Another feature of the plane's design was a .30-caliber Browning machine gun that was synchronized to fire through the plane's propeller. Thus, the TS-1 can be called the first Navy plane actually planned as a fighter.

The Naval Aircraft Factory built five TS-1s, and a contract for 34 more planes was awarded to the Curtiss Aeroplane and Motor Company. The first TS-1s were delivered to the *Langley* in December, 1922.

The TS-1 was sometimes tricky to put down on the *Langley*'s deck because of its springy landing gear, but this drawback was offset by the plane's low landing speed. A dependable aircraft, the TS-1 remained in service until 1926.

The TS-1 did not serve only as a carrier-based aircraft. Equipped with floats instead of wheels, it also operated with destroyers, cruisers, or battleships. In such cases, the aircraft would be slung over the side from a crane to take off from the water, then hoisted back aboard for storage.

Other Data (Model: TS-1)
Wingspan: 25 ft.
Length: 22 ft., 1 in.
Power Plant: One 200-hp Wright J-4
Loaded Weight: 2,133 lb.
Maximum Speed: 123 mph at sea level

Planes of the FB series were light in weight and highly maneuverable, and very popular with Navy pilots of the 1920s.

BOEING FB

In 1920, Boeing was a small Seattle, Washington, aircraft company building wooden Thomas Morse MB-3A fighters for the Army Air Corps. The contract kept the skilled work force busy and production lines humming.

But Boeing's engineers felt that they could do better. They had made a careful study of the German Fokker D-7, a number of which had been captured during World War I and brought to the United States. The German fighter was very strong but light in weight, thanks to a frame made of welded steel tubing that was fabric-covered.

Boeing decided to produce a single-seat, high-performance fighter, also using welded steel tubing. Boeing's engineers were confident that they could turn out a better plane than the Fokker because they planned to use newly perfected electric arc-welding techniques. The Germans had used gas welding.

The result of Boeing's efforts was an aircraft designated Model 15. It was first flown in the spring of 1923. So superior was the Model 15 to existing fighters that the Army immediately placed an order for 16 planes.

An aircraft similar to the Model 15 was ordered by the Navy in 1925. Designated the FB-1, the aircraft was not equipped for carrier operations. It had to be modified by strengthening the fuselage and undercarriage and by installing axle hooks to be used for landings.

The first true carrier version of the plane was the FB-5, delivered to the Navy in 1927. Twenty-eight FB-5s were assigned to the USS *Langley*. Their very first flights were takeoffs from the carrier at sea.

The plane proved to be very popular with the pilots. It was extremely maneuverable and drifted in for landings at speeds of only 60 mph.

The plane's armament consisted of either twin .30-caliber machine guns or one .30- and one .50-caliber gun. In either case, the guns were synchronized to fire through the arc of the propeller.

Although the planes in the FB series were never ordered in great quantity, they helped to establish the Boeing Company as a leader in fighter plane design and development.

Other Data (Model: FB-1)
Wingspan: 32 ft.
Length: 23 ft., 5 in.
Power Plant: One 435-hp Curtiss D-12
Loaded Weight: 2,835 lb.
Maximum Speed: 159 mph

Like the Navy planes that had come before, aircraft in the FB series were built as both land and seaplanes.

CURTISS F6C HAWK

In 1921, the Curtiss Aeroplane and Motor Company began to develop racing planes for both the Navy and the Army Air Service. The two services competed against one another in the Pulitzer Prize Races, which were popular at the time. Several Curtiss planes established world speed records. As a result, Curtiss decided to adapt their basic racing-plane design for use as a single-seat fighter.

The Army Air Service ordered the first airplane in the series in 1925. It was designated the PW-8. When the Air Service changed its designation system, the PW-8 became the P-1. The Navy ordered nine P-1s, also in 1925, and called them F6C-1s.

These were the first in a series of Hawk fighters purchased by the Navy. The next version, the F6C-2, was equipped with arrester hooks and strengthened landing gear for carrier operations.

The same features were included in the 35 F6C-3s ordered by the Navy in 1927. These aircraft were flown by a squadron nicknamed the Red

The Navy ordered thirty-five F6C-3s in 1927. They were flown by a squadron nicknamed the Red Rippers.

A number of F6C-3s saw service with the Marine Corps and were based at Quantico, Virginia.

Rippers that specialized in aerial acrobatics. A number of other F6C-3s saw service with a Marine Corps squadron at Quantico, Virginia.

Each of the first three versions of the F6C were powered by a Curtiss liquid-cooled engine. But the Navy complained that these engines were difficult to service. For the F6C-4, a 410-hp Pratt & Whitney air-cooled Wasp was substituted for the Curtiss engine.

The Wasp was an engine of radial design. Its pistons moved inward and outward from a central shaft. Not only did it prove easier to maintain and service, but the new engine delivered greater power than the one it replaced. From that time on, with only a small handful of exceptions, the Navy specified radial engines for its piston-powered aircraft.

Despite the new engines, the F6C-4s were all but out-of-date at the time they were delivered. Used only aboard the *Langley*, they were withdrawn from service in 1930.

Other Data (Model: F6C-4)
Wingspan: 37 ft., 6 in.
Length: 22 ft., 6 in.
Power Plant: One 410-hp Pratt & Whitney Wasp
Loaded Weight: 3,171 lb.
Maximum Speed: 155 mph at sea level

The FU-1 failed in its role as a carrier plane because of limited cockpit visibility.

VOUGHT FU-1

One of the biggest problems naval officials faced in attempting to marry the airplane with ships of the fleet was how to get each aircraft into the air and back on board again without disrupting shipboard operations. The catapult provided at least part of the solution.

The Navy tested a primitive catapult in 1912. It consisted of a track mounted on a long barge. Using compressed air, a heavy cart with small wheels would be propelled along the track at high speed.

The next step was to load an airplane onto the cart and rev up its engine. At a signal, the cart would be sent rocketing down the track. At the end of its run, the aircraft, having reached takeoff speed, would lift into the air.

For years, battleships and cruisers carried aircraft and launched them from deck catapults of this type. Early aircraft designed for catapult launching were float planes. When an aircraft had completed its mission, it would land in the water beside its mother ship and a crane would lift it aboard.

On June 30, 1926, the Navy placed an order for 20 single-seat catapult-launched float-plane fighters to serve aboard battleships. These were not to be newly designed aircraft, but merely modified versions of the Vought UO-1, which, in turn, was an improved model of the Vought VE-7 (page 6). The plane's armament was to consist of a pair of .30-caliber Browning machine guns.

The new FU-1s, as they were called, were distributed among twelve ships of the battle fleet. After eight months, however, they were removed from service with battleships and converted to carrier operations. This meant replacing the floats with conventional landing gear.

Carrier pilots complained about the poor visibility from the FU-1's cockpit, which was a good distance from the plane's nose. The solution was to open up a forward cockpit position. Once the FU-1 became a two-place aircraft, it was no longer considered suitable for use as a fighter. Many of the aircraft were reassigned to duties as trainers.

Other Data (Model: FU-1)
Wingspan: 34 ft., 4 in.
Length: 28 ft., 4½ in.
Power Plant: 220 hp-Wright J-5
Loaded Weight: 2,744 lb.
Maximum Speed: 122 mph at sea level

When a forward position was opened up in its fuselage, the FU-1 became a two-place trainer.

BOEING F3B

During the late 1920s, it was standard practice for the Navy's two biggest aircraft carriers, the *Saratoga* and *Lexington*, to carry four air squadrons. Each squadron had eighteen planes. One squadron was made up of fighters, a second of scout planes, a third of torpedo-dropping aircraft, and the fourth of dive bombers.

Since this was a period before any aircraft had been designed specifically for dropping bombs while diving at the enemy, the role of the dive bomber on both carriers was handled by a fighter plane—the Boeing F3B.

The F3B was a modified version of the F2B which, in turn, had been developed from the FB (page 10). Pilots liked the F2B, just as they had the earlier FB. But the F3B was not nearly as popular. It was a heavier plane than its predecessors, had a longer fuselage and greater wingspan, and did not handle as easily.

The F3B-1 was first flown on February 3, 1928.

Heavier than the planes that preceded it, and with a longer fuselage and greater wingspan, the F3B was not an easy plane to handle.

F3Bs, their engines cloaked in protective covers, crowd a carrier flight deck. (Other aircraft pictured are F6Cs.)

Seventy-four production models were ordered by the Navy. Deliveries began in August, 1928.

The F3B was the first American fighter to have all-metal tail surfaces and ailerons. (Ailerons are the movable surfaces at the trailing edges of the wings that control banks and rolls.) The plane was armed with one .30-caliber and one .50-caliber machine gun, both forward firing.

Pilots of F3Bs, in playing the dive-bombing role, developed a tactic called the "split dive." The eighteen aircraft of the squadron would divide into three groups. The planes in each group would dive in sequence at a common target, each group diving from a different direction. Only aerial exhibition teams employ such maneuvers today.

Fighters continued to operate as dive bombers until the end of the 1920s. By the beginning of the next decade, they were beginning to be replaced by the Martin BM and other aircraft developed specifically to be dive bombers.

The F3B was not needed as a fighter, either. That role was taken over by Boeing's F4B, which was to become one of the most famous Navy fighters of all time.

Other Data (Model: F3B-1)
Wingspan: 33 ft.
Length: 24 ft., 10 in.
Power Plant: One 425-hp Pratt & Whitney Wasp
Loaded Weight: 2,945 lb.
Maximum Speed: 157 mph at sea level

Lighter than the F3B, and equipped with a more powerful engine, the F4B was a joy to fly.

BOEING F4B

In 1928, when the Boeing Company was called upon to develop a high-performance carrier-based fighter, a plane more compact and maneuverable than any of its predecessors, the company did just that. There was nothing radically new or different about the aircraft that resulted, except that it was better in almost every way than those it was meant to replace.

Boeing engineers reduced weight by replacing welded steel tubing in the fuselage with duralumium that imparted strength as well as lightness.

They reduced the wing by almost 15 square feet from what it had been in the F3B. They shortened the fuselage by four feet. As a result, the F4B, as the new aircraft was called, was lighter than the F3B. When a slightly more powerful engine was added, greater speed was the result. Pilots found the F4B to be a joy to fly.

The first F4B-1s, an order of 27, were delivered to fighter squadrons aboard the *Langley* and *Lexington* in May and June of 1929.

Boeing made a major change in 1931 by designing the plane with an all-metal fuselage. In addition, the aircraft was given a more powerful engine and bigger tail fin and rudder. The Navy contracted for 75 of this plane, the F4B-3.

The final version of the aircraft was the F4B-4, introduced in 1932. It had wing bomb racks and larger tail surfaces than the F4B-3.

F4B-4s remained in service with the Navy until 1937. By the middle of 1938, they had been replaced by speedier biplanes manufactured by the Grumman Company.

But the F4B-4 continued to serve in a utility role until the 1940s. A total of 586 planes of this type were built by Boeing. Less than half of these went to the Navy, however. The U.S. Army Air Force ordered almost 350 of them. To the Army, the F4B was known as the P-12.

It is believed that only one F4B ever saw combat. Sold to China, the aircraft became caught up in the Sino-Japanese war that erupted in 1932. The F4B was downed by Japanese planes after accounting for two of its three attackers.

F4Bs, bristling with wing-mounted .30-caliber machine guns, fly in formation.

Other Data (Model: F4B-1)
Wingspan: 30 ft.
Length: 20 ft., 1 in.
Power Plant: One 450-hp Pratt & Whitney Wasp
Loaded Weight: 2,750 lb.
Maximum Speed: 176 mph at 6,000 ft.

CURTISS F9C SPARROWHAWK

In 1926, the Navy was given the go-ahead to obtain two huge rigid airships for scouting purposes. The Goodyear-Zepplin Company of Akron, Ohio, was assigned to produce them. Named the *Akron* and *Macon*, the two airships began operations in 1931.

Because the airships moved at painfully slow speeds, the Navy realized that they would be easy prey for enemy aircraft. The best way to protect the dirigibles, the Navy decided, was to equip each of them with fighter aircraft.

What was needed was a very small, very light plane. Earlier, in the spring of 1930, the Navy had asked several aircraft companies to submit designs for a small carrier-based fighter. One of these planes, the Curtiss XF9C-1, proved to be what the Navy needed to guard its airships.

Eight F9C Sparrowhawks were built, including two experimental models. The six production aircraft were delivered to the Navy in September, 1932, and began operating from the airship *Akron*.

Each plane had a "skyhook" mounted to its fuselage above the wing and behind the engine. When his mission was completed, the pilot would slow his plane's speed to exactly that of the airship. He would then guide the hook to the horizontal bar of a recovery trapeze that hung from beneath the airship.

Once hooked onto the bar, the pilot would cut the plane's engine. The trapeze would then be raised into the open belly of the airship. There the plane would be stored. The precedure was reversed when launching a plane.

Tragedy marred the Navy's airship program. The *Akron* was lost in a storm at sea in 1933. A similar mishap claimed the *Macon* in 1935.

The orphaned Sparrowhawks remained in service as utility planes for another year or two. One is now on display at the National Air and Space Museum in Washington, D.C., its "Man on the Flying Trapeze" insignia recalling one of naval aviation's most colorful periods.

Other Data (Model: F9C-2)
Wingspan: 25 ft., 5 in.
Length: 20 ft., 7 in.
Power Plant: One 438-hp Wright Whirlwind
Loaded Weight: 2,770 lb.
Maximum Speed: 176 mph

Big skyhook mounted to the Sparrowhawk's fuselage above the wing enabled the aircraft to "hook up" to an airship.

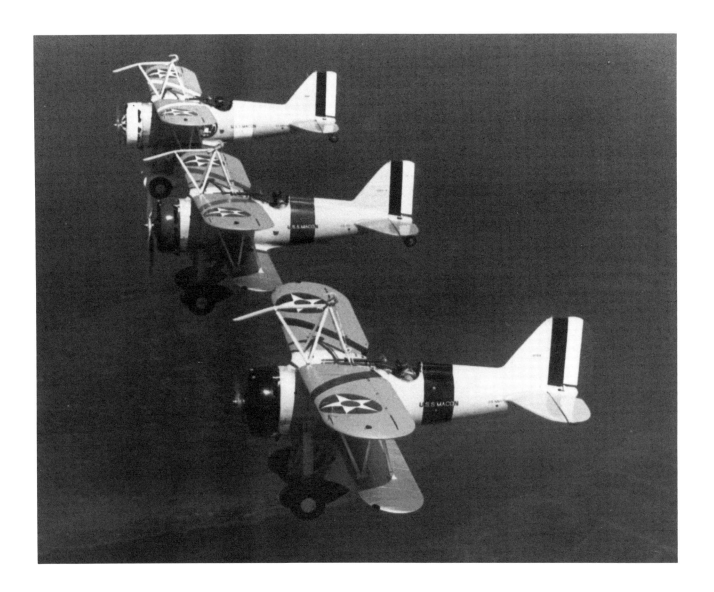

The FF-1 introduced the retractable landing gear to Navy fighters.

GRUMMAN FF

The Grumman FF, or "FiFi," as it was nicknamed, was the first Navy fighter to offer a retractable landing gear. The plane's front wheels folded upward into circular wells on each side of the fuselage just in front of the lower wing.

The FF had a fuselage with a metal skin and fabric-covered wings. Although it was a two-seater, it was faster than many of the Navy's single-seat fighters.

The plane was built by Leroy Grumman, who started his aircraft company in a garage in Baldwin, Long Island, New York. Grumman was making floats for Navy seaplanes when, in 1931, he received a contract to develop the prototype for a two-seat fighter. The aircraft, first flown toward the end of 1931, was designated the XFF-1.

The new aircraft was armed with a .30-caliber forward-firing machine gun in the front cockpit, and two Brownings of the same size in the rear cockpit. The plane was easily recognizable because of the long canopy that covered the two cockpits.

The XFF-1 was an immediate success. Powered

by a 616-hp Wright Cyclone engine, it achieved 195 mph during test flights. Later, after a more powerful engine had been installed, the aircraft surpassed 200 mph.

The Navy ordered 27 FF-1s. Deliveries began in June, 1933. The aircraft were assigned to squadrons aboard the *Lexington*. A number were also based in Canada and operated by the Royal Canadian Air Force. There the plane was known as the Goblin I.

Grumman also developed a modified version of the FF-1 to be used for scouting duties. Thirty-three of these were produced. Like the fighters, they served aboard the *Lexington*.

All FF-1s had been withdrawn from service by 1936, and were handed over to units of the Naval Reserve.

The FF-1 signaled the beginning of the Grumman Company's long relationship with the Navy. In the decades that followed, it would produce some of the Navy's most successful aircraft.

Other Data (Model: FF-2)
Wingspan: 34 ft., 6 in.
Length: 24 ft., 6 in.
Power Plant: One 700-hp Wright Cyclone
Loaded Weight: 4,828 lb.
Maximum Speed: 207 mph at 4,000 ft.

A squadron of FF-1s fly in formation high above the carrier *Saratoga*.

GRUMMAN F2F, F3F

The successful flights of the XFF-1 led the Navy to order a single-seat version of the plane. Designated the XF2F-1, it had the same stubby appearance of its predecessor. Pilots called it the "Flying Barrel."

The XF2F-1 was first flown on October 18, 1933. Like the FF-1, the aircraft had an enclosed cockpit, the first Navy single-seater to offer this benefit.

In test flights, the prototype traveled at a top speed of 229 mph at 800 feet. It could climb at a rate of more than 3,000 feet per minute. Impressed, the Navy ordered 54 F2F-1s.

By mid-1938, squadrons aboard the *Ranger* and *Lexington* were flying the new biplane. When a squadron from the *Lexington* was reassigned to the *Yorktown* and later to the *Wasp*, the pilots took their F2Fs with them.

Although the plane was serving with carrier fighter squadrons throughout the fleet, the Navy was still tinkering with the F2F's design. Grumman was asked to produce a prototype with a longer fuselage and increased wingspan, modifications

The plump, single-seat F2F, nicknamed the "Flying Barrel," served with carrier squadrons throughout the fleet.

that were intended to make the plane more maneuverable.

The new plane, designated the XF3F-1 crashed during flight testing in May, 1935, but was soon replaced. The test flights that followed pleased Navy officials.

A total of 164 F3Fs were eventually ordered by the Navy. The first of these reached the fleet in 1936, supplying squadrons aboard the *Ranger* and *Saratoga*.

A later version of the plane, designated the F3F-3, tested in January, 1937, boasted a 950-hp Wright Cyclone engine. This plane could travel at much faster speeds than any of the earlier models, reaching 264 mph. The first productions of the F3F-3s were assigned to the *Yorktown*.

Grumman F2Fs and F3Fs remained in service with the fleet until the early months of 1941. They were then handed out to naval air stations where they were assigned to run errands.

The F2Fs and F3Fs are notable because they were the last of the fighter biplanes to fly with the U.S. Navy, indeed, with any branch of the armed services. With their retirement, an era ended.

Other Data (Model: F2F-1)
Wingspan: 28 ft., 6 in.
Length: 21 ft., 6 in.
Power Plant: One 700-hp Pratt & Whitney Twin Wasp
Loaded Weight: 3,847 lb.
Maximum Speed: 231 mph at 7,500 ft.

Other Data (Model: F3F-3)
Wingspan: 32 ft.
Length: 23 ft., 2 in.
Power Plant: One 950-hp Wright Cyclone
Loaded Weight: 4,795 lb.
Maximum Speed: 264 mph at 15,200 ft.

The F3F was a stretched-out version of the F2F, and it had a greater wingspan.

Although it never won high marks for its performance, the F2A stands as the Navy's first monoplane fighter.

BREWSTER F2A BUFFALO

By the mid-1930s, American aircraft designers were aware that the biplane fighter was fast going the way of the horse and wagon. It was not possible to improve existing biplane designs to any degree. The future belonged to the single-wing plane, or monoplane, which was speedier.

In 1936, the Grumman Company produced a prototype for a monoplane fighter, the XF4F-2. (The XF4F-1 had been a biplane.) The plane was flight-tested the following year.

The Brewster Aeronautical Corporation of Johnstown, Pennsylvania, also produced a monoplane prototype. Brewster had more experience with monoplanes than Grumman, having built scout bombers of that design.

The Brewster monoplane fighter, the XF2A, bore a family resemblance to the company's bomber. It was a short and stubby aircraft with a mid-mounted wing. The aircraft's main wheels folded inward, that is, toward one another, to be stored in wheel wells in the plane's fuselage.

The XF2A-1 was flown for the first time in

December, 1937. Although flight tests demonstrated that the aircraft was not as fast as the monoplane that Grumman was developing, and there were problems with cockpit visibility, the Brewster plane was ordered into production. The first models were delivered during June, 1940.

F2A-1s were supplied to Belgium and England early in World War II. In 1941, when Italian and German troops attacked Crete, F2A-1s from British carriers were used in the island's defense. But the plane proved to be underpowered, and was a disappointment.

The Brewster Company tried to improve the F2A. A more powerful engine was installed. Armor protection for the pilot was added. The bigger engine should have resulted in greater power and speed, but the weight of the armor plate caused even poorer performance.

Only a handful of F2A-3s served with the U.S. Navy. The plane had a sad combat record against Japanese fighters. Happily, its career was brief.

Other Data (Model: F2A-3)
Wingspan: 35 ft.
Length: 26 ft., 4 in.
Power Plant: One 1,200-hp Wright Cyclone
Loaded Weight: 6,840 lb.
Maximum Speed: 313 mph at 13,500 ft.

At 13,500 feet, the F2A-3 could reach a speed of 313 mph—but its rival, the F4F, could do better.

GRUMMAN F4F WILDCAT

On December 7, 1941, when the Japanese attacked Pearl Harbor and the United States was plunged into World War II, the Grumman F4F Wildcat was the Navy's No. 1 fighter. A small, single-engine, midwing monoplane, the Wildcat was rugged and reliable and, in the hands of a skilled pilot, able to hold its own with Japan's deadly Zero fighter. More than 7,900 Wildcats were built.

The first plane designed in a series of planes that was to become the F4F was a biplane. Designated the XF4F-1, it was based upon an earlier Grumman fighter, the chunky little F3F (page 24).

The biplane design was soon discarded in favor of an all-metal monoplane, the XF4F-2. This plane had to compete with the Brewster F2A Buffalo for the Navy's blessings—and the F2A won out.

But since the Grumman plane had shown promise, the Navy asked the company to rework the design. The result was the XF4F-3, a greatly improved aircraft, one that was clearly superior to the Brewster Buffalo. Grumman was given a production order for 54 F4F-3s in August, 1938.

In spite of the high hopes for the Wildcat, the plane was often severely mauled in the early stages of World War II. At Pearl Harbor, eleven Wildcats were caught on the ground by Japanese bombers. Nine were destroyed.

During the early stages of World War II, the F4F was the Navy's chief carrier-based fighter.

At Wake Island, seven Wildcats were wiped out during the first phase of the Japanese attack. But the small number of remaining planes was successful in downing a twin-engine Japanese bomber and a Zero in air combat, and a Marine Corps Wildcat pilot sank an enemy destroyer before Wake Island was overrun.

The Wildcat's chief opponent was the Japanese Zero, a plane that could outperform and outmaneuver it. The Wildcat's strong points were its heavy armament and solid construction. It could absorb punishment and carry on. The Zero could not.

In aerial combat with the Zero, Wildcat pilots tried to achieve head-on or diving attacks. In a head-to-head duel, the Wildcat had the advantage of the superior firepower of its high velocity .50-caliber machine guns. The Zero had smaller, slow-firing 7.7-mm guns, less than .30-caliber in size.

Lt. Edward H. (Butch) O'Hare, one of the first American heroes of World War II, was a Wildcat pilot. On February 20, 1942, he and other members of his squadron from the *Lexington* came upon a large force of Japanese bombers returning to their base after a raid. In the battle that followed, O'Hare shot down five planes and damaged a sixth. He became one of the first American aces of World War II. (An ace is a pilot who has downed at least five enemy aircraft.) O'Hare also received the Medal of Honor.

Flight deck crew aboard an escort carrier readies an F4F for launching.

Wildcats played important roles at the Battle of the Coral Sea and at Midway. Not only was the plane on constant duty in the Pacific, but the Wildcat squadrons also aided in the Allied invasion of French North Africa in October, 1942.

The F4F went through several modifications. The original four .50-caliber guns were increased to six. Then the number was reduced to four again. Pilots said six-gun firepower was more than was

Flight deck of the carrier *Saratoga* in October, 1941, featured F4F Wildcats.

required to bring down the lightly protected Japanese fighters.

The plane's Pratt & Whitney engine was upgraded with the addition of a two-stage, two-speed supercharger. The solid wing in the F4F-3 became a folding wing in the F4F-4. This modification greatly increased the number of aircraft that could be stored aboard a carrier.

As early as 1941, Grumman had begun the development of an aircraft that would be superior to the Wildcat and eventually replace it. That plane, the F6F Hellcat, became available in large numbers in 1943.

To allow Grumman to concentrate on the Hellcat, production of the Wildcat was shifted from its Long Island, New York, plant to the Eastern Aircraft Division of General Motors in Trenton, New Jersey. Wildcats turned out by General Motors were designated as FM-1s and, later, FM-2s.

The FM-2 had a more powerful engine and a lighter airframe than previous versions of the aircraft. With these changes, the plane was well suited for use aboard the smaller escort carriers with their shorter flight decks. The Wildcat became standard equipment on the majority of the Navy's 114 escort carriers.

Through the first half of the war, the F4F served heroically as the Navy's only carrier-based fighter. Although replaced by the F6F Hellcat and F4U Corsair as the Navy's frontline fighter in the Pacific, it continued to operate with escort carriers, serving as an antisubmarine fighter in the Atlantic and supporting troop landings in the Pacific.

Other Data (Model: F4F-4)
Wingspan: 38 ft.
Length: 28 ft., 9 in.
Power Plant: One 1,200-hp Pratt & Whitney Twin Wasp
Loaded Weight: 7,952 lb.
Maximum Speed: 318 mph at 19,400 ft.

F4F at the right is piloted by Lt. Edward H. (Butch) O'Hare, one of the first American heroes of World War II.

GRUMMAN F6F HELLCAT

To be successful as a fighter during World War II, an aircraft had to have superior speed, be able to climb faster than its rivals, and boast greater overall maneuverability.

In the case of a Navy plane operating at sea, it also had to be rugged enough to withstand the abuse of carrier launchings and landings.

When World War II began in the Pacific, Japan had the best shipboard fighter in the Mitsubishi A6M2, better known as the Zero.

F6F Hellcat has been hailed as the finest carrier-based aircraft to see action during World War II.

The Zero was a big shock to Allied forces, scoring one success after another. Pearl Harbor and the attack on Wake Island were only the beginning. Zeros pounded Port Darwin, Australia, in February, 1942, and then headed for the Indian Ocean where they sank several British naval vessels.

Even at the Battle of Midway, considered a big victory for American forces, Zeros made their presence known, attacking American torpedo bombers and inflicting terrible losses.

Why was the Zero so successful? It was a light plane with a very powerful engine. That was its secret. Japanese designers sacrificed everything to achieve speed and maneuverability.

It was a plane built to attack. It had no armor plating to protect the pilot or fuel tanks. It could not absorb punishment. One accurate burst of fire and the Zero would usually erupt in flames.

Because the plane was so maneuverable, Navy pilots were not to engage in dogfights with the Zero. Instead, they were instructed to dive on the Zero from above, get in a short burst of fire and then zoom up to regain altitude.

The F4F Wildcat was adequate as a Navy fighter. Adequate means barely sufficient. Something better was needed, a plane that could clearly outduel the Zero.

Designers at Grumman talked to Navy pilots with combat experience and asked them for their

Fast and rugged, the **F6F Hellcat** enabled U.S. forces to gain control of the skies in the Pacific during World War II.

Tailhook down, an F6F drifts in for carrier landing.

views. The result was a fighter that was faster than the Zero and had excellent flight control characteristics. It also had the ruggedness of the Wildcat.

Designated the F6F Hellcat, the new plane enabled United States forces to take control of the skies in the Pacific. The Hellcat has been called the finest carrier-based aircraft to see service during World War II.

At first glance, the Hellcat looked very much like the F4F Wildcat. But on closer inspection, one could spot the differences. Instead of having a mid-wing layout, the Hellcat was of low-wing design. This made it possible for the main wheels to be retracted into the wing. In the Wildcat, they folded into the fuselage.

The wheels in the Hellcat were also positioned farther apart. This gave the plane greater stability on takeoffs and landings. Other improvements included more armor protection for the pilot and increased storage space for ammunition.

The first flight of the XF6F-3 took place in July, 1942. Within six months, the first Hellcats were ready for delivery.

Hellcat squadrons went into action for the first time on August 31, 1943, attacking the Japanese-held Marcus Islands in the central Pacific. Enemy pilots first thought the planes were Wildcats and were confident that they had them outmatched. They quickly discovered they had encountered fighters that were much superior to their own.

Hellcats were a major factor during the Battle of the Philippine Sea in January, 1944, and the Battle of Leyte Gulf later in the year, where the Japanese suffered staggering losses in planes and ships. During these engagements, it was the responsibility of the Hellcats to defend the carrier strike forces from the suicidal attacks of the Japanese pilots.

A total of 11,013 Hellcats were built for the Navy. The plane was second only to the Corsair (page 36) in terms of number of aircraft produced.

The Hellcat was also widely used by the British Navy's Fleet Air Arm. Some 1,182 F6Fs were exported to Great Britain. These planes were used to attack enemy shipping along the Norwegian coast and to provide fighter cover during attacks on the German battleship *Tirpitz* during 1944.

Hellcat production ended in November, 1945. For a plane that was used so extensively, it was modified very little. It was designed to do a specific job: win air superiority in the Pacific. And that's exactly what it did. There was never any need to change anything.

Other Data (Model: F6F-5)
Wingspan: 42 ft., 10 in.
Length: 33 ft., 7 in.
Power Plant: One 2,000-hp Pratt & Whitney Double Wasp
Loaded Weight: 15,413 lb.
Maximum Speed: 380 mph at 23,400 ft.

Although no Hellcats were produced after World War II, the fighter continued to serve with Navy Reserve squadrons.

Its inverted gull wing gave the F4U Corsair a distinctive appearance.

VOUGHT F4U CORSAIR

A big fighter of unusual design, the F4U Corsair overcame early development problems to eventually rival the F6F Hellcat as the best carrier-based aircraft of World War II. In one respect, it had no rivals. More Corsairs were built—12,620 of them—than any other World War II fighter plane.

The Corsair also had the distinction of remaining in continuous production longer than any other fighter. Designed in 1938, and first flown during May, 1940, the Corsair was still being produced as late as 1952. One F4U was on active duty with the service until 1962.

In developing the Corsair, designers began with the idea of combining the smallest possible airframe with the most powerful engine then available.

To take advantage of the engine's exceptional power at high altitudes, a very long propeller was required. The big propeller created a problem. It had such a wide turning arc that its tips would scrape the ground unless the fuselage was elevated by a tall landing gear.

This caused another problem. A tall landing gear had a tendency to collapse during carrier landings. Carrier-based planes required a landing gear that was short and rugged.

These problems were solved by fitting the plane with bent wings. (Technically, the design was known as an inverted gull wing.) The landing gear was then positioned beneath the point where the bend was located, which was closer to the ground than any other part of the wing. This enabled designers to keep the Corsair's landing gear as short as possible.

The Corsair prototype, the XF4U-1, was flown for the first time on May 29, 1940. During its testing, the aircraft startled observers by reaching a speed of 404 mph, faster than any other fighter then in the air.

Still, all was not well with the plane. When production models began to be delivered in 1942, they caused some anxious moments. In approaching the carrier deck, the Corsair would sometimes stall without any warning. And when the plane did reach the flight deck, it bounced heavily, sometimes causing the tires to burst or damaging the fuselage.

The Navy decided that the plane was unsuited for carrier operations until the problems could be corrected. The first Corsairs were handed over to land-based Marine Corps units.

Corsairs first saw combat duty at Guadalcanal,

Land-based F4Us, their wings still folded, prepare for takeoff.

one of the Solomon Islands, in February, 1943. It was during the struggle for control of Guadalcanal that Maj. Gregory "Pappy" Boyington, the leading

A total of 12,620 Corsairs were built during World War II. No other Navy aircraft was produced in that quantity.

Marine ace of World War II, began his brilliant record. Boyington was a Corsair pilot.

To get as many Corsairs into production as quickly as possible, Goodyear Aircraft Corporation and Brewster Aeronautical Corporation were called upon to set up production lines. The Vought Company remained the principal manufacturer, of course.

A total of 2,012 Corsairs were shipped to Great Britain during World War II. The first of these arrived during 1943. A British Corsair squadron went into action on April 3, 1944, operating from the carrier *Victorious*. At about the same time, the Corsair finally won approval from the Navy for use aboard American carriers.

The Corsair proved to be a very versatile plane. On some, long-range fuel tanks were installed beneath the fuselage and two 1,000-pound bombs or eight 5-inch rockets were placed beneath the wings. These Corsairs were fighter-bombers.

Other Corsairs were fitted with wingtip radar and used as night fighters. Night-flying Corsairs operated from the *Essex*, *Hornet*, and *Intrepid*. Still other Corsairs were packed with camera equipment. They were reconnaissance aircraft.

One Corsair was responsible for one of the most remarkable kills of World War II. It took place during air combat in the skies over Okinawa late in the war. Lt. Robert T. Kingman was the pilot.

Kingman chased a Japanese Nick to 38,000 feet,

F4U Corsairs on the flight deck of an aircraft carrier just before the Inchon invasion in Korea in 1950.

whereupon his guns jammed and stopped firing. Kingman brought down the enemy plane by clipping off its rudder and other chunks of the tail assembly with the Corsair's propeller.

By the end of World War II, Corsairs had compiled a remarkable record. They had shot down 2,140 enemy planes, while losing only 189, a victory ratio of slightly better than eleven to one.

During the Korean War in the early 1950s, Corsairs scored ten aerial combat victories. One Corsair was even credited with bringing down a Russian-built MiG-15. It was the last known victory of a piston-powered plane over a jet.

Other Data (Model: F4U-4)
Wingspan: 41 ft.
Length: 33 ft., 8 in.
Power Plant: One 2,100-hp Pratt & Whitney
   Double Wasp
Loaded Weight: 14,670 lb.
Maximum Speed: 446 mph at 26,200 ft.

GRUMMAN F7F TIGERCAT

A twin-engine, carrier-based fighter, the F7F Tigercat arrived too late to see service in World War II. The big plane was intended for use aboard the huge "battle carriers" of the *Midway* class. The first of these was not completed until 1945, the year the war ended.

The F7F Tigercat had its beginnings in 1938, the year the Navy asked Grumman to develop the XF5F Skyrocket, a twin-engine plane. When test-flown, problems developed in cooling the Skyrocket's engines. And the engine housings were so big, the pilot could not get a clear view of the carrier flight deck when landing. The Skyrocket never went into production.

F7F Tigercat was the Navy's first successful twin-engine carrier-based fighter.

Firepower provided by four .50-caliber machine guns and four 20-mm cannons made the Tigercat a potent ground support weapon.

But Grumman was able to draw upon its experience with the Skyrocket when the company was asked to develop the twin-engine Tigercat. The first prototype for the new plane made its first flight in December, 1943.

Besides operating as a fighter, the F7F was meant to play a ground-support role, and for this it had to be very heavily armed. There were four .50-caliber machine guns in the nose and four 20-mm cannons in the wings. In addition, the Tigercat could carry a torpedo or a 1,000-pound bomb beneath its fuselage. In 1945, the Navy could rightly call the Tigercat "the most powerful fighter-bomber in the world today."

The first F7F-1s produced by Grumman were single-seat aircraft. But later a two-seat Tigercat was developed. This was a night fighter. A radar operator occupied the second seat. Some camera-carrying Tigercats were also produced to serve as reconnaissance planes.

While some 364 F7Fs were built for the Navy, none ever served in World War II. The Tigercat did see action during the Korean War, however. Operated by Marine squadrons, the powerful and heavily armed aircraft was particularly effective as a night fighter in supporting ground operations.

Other Data (Model: F7F-3)
Wingspan: 51 ft., 6 in.
Length: 45 ft., 4 in.
Power Plant: Two 2,100-hp Pratt & Whitney
   Double Wasps
Loaded Weight: 25,720 lb.
Maximum Speed: 435 mph at 22,000 ft.

Tall tail fin was a distinctive feature of the F8F Bearcat. It also had a smaller airframe than earlier Grumman fighters.

GRUMMAN F8F BEARCAT

Considered by many to be the Navy's best piston-engined fighter, the F8F Bearcat represented the final development in the line of Grumman fighters that started with the FF (page 22). Although the plane never flew in combat for the Navy, it did see service with the Royal Thai Air Force and in Vietnam with the French *Armee de l'Air*.

The F8F Bearcat was originally planned as an improved version of the F6F Hellcat. While the overall appearance of the Bearcat was similar to that of the Hellcat, the airframe was smaller and the wingspan shorter. The Bearcat was also equipped with a more powerful engine. The first Bearcat flew on August 31, 1944.

When the plane was tested, the Navy called the results "spectacular." Its speed of more than 400 mph at sea level was believed to be the world's fastest for a propeller-driven aircraft. Within five months after the Bearcat's first flight, the first production Bearcats were rolling off the Grumman assembly line.

The Navy ordered more than 2,000 F8F-1s from Grumman and 1,876 from the aircraft division of General Motors. The first deliveries of the production models of the plane began in February, 1945.

World War II ended with VJ Day, August 14, 1945. The Navy drastically cut back the number of Bearcats on order. None of the General Motors planes were ever built. Grumman turned out a total of 1,263 Bearcats.

This number included 100 planes produced after the war. These were armed with four 20-mm cannons, replacing the original armament of four .50-caliber machine guns. Night-fighting Bearcats were also produced as well as camera-bearing aircraft for reconnaissance duty.

Had World War II continued, observers agree that the Bearcat would have come to rank with its predecessor, the F6F Hellcat, and the Vought F4U Corsair as one of the Navy's standout fighter planes.

Other Data (Model: F8F Bearcat)
Wingspan: 35 ft., 10 in.
Length: 28 ft., 3 in.
Power Plant: One 2,400-hp Pratt & Whitney Double Wasp
Loaded Weight: 12,947 lb.
Maximum Speed: 421 mph at 19,700 ft.

Had World War II continued, the Bearcat might have been ranked with the Hellcat and Corsair as one of the Navy's great fighters.

When making a carrier landing or takeoff, the Fireball relied on its piston engine.

RYAN FR FIREBALL

By the end of 1942, the obvious superiority of jet-propelled aircraft was well known. The German air force had flown a jet fighter, the Messerschmitt Me. 262 Schwalbe (swallow) on June 18, 1942. By 1943, the U.S. Air Force would have a prototype for its first jet, the F-80 Shooting Star, in the air.

When the Navy's engineers began experimenting with this new type of propulsion, they faced a serious problem. Jet aircraft of the day were so slow in accelerating that carrier takeoffs were out of the question. There was no flight deck in the world long enough to accommodate them.

The Navy's solution was to suggest a single-seat fighter with both a piston engine and a jet engine. The aircraft would use the piston engine when taking off and for landings, too. The jet was called upon when the plane became airborne and additional speed was needed.

Nine manufacturers were asked to prepare designs for such an aircraft. The Ryan Aeronautical Company, known for the famous "Spirit of St. Louis," the plane that Charles Lindbergh had used in making the first solo transatlantic flight in 1927, submitted the winning proposal. Three prototypes were ordered.

The XFR-1, as the aircraft was called, made its first flight on June 25, 1944. A simple low-wing monoplane with a tricycle landing gear, the aircraft had its piston engine in the nose and the jet engine at the rear of the fuselage. The first production models of the unusual aircraft were delivered in January, 1945.

Test landings and takeoffs were conducted aboard the carrier *Ranger* in May, 1945. The Fireball, a name given the plane by the builder, thus became the Navy's first operational jet aircraft.

When World War II ended in August, 1945, production of the Fireball ceased. Only 66 of the jet/prop planes were built. It is as an experimental aircraft that the Fireball is remembered.

Once airborne, the Fireball cruised on its jet. The piston engine could be turned off (as indicated by non-turning propellers in this photo).

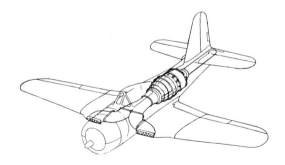

Fireball's piston engine was positioned at the aircraft's nose; its jet, within the fuselage.

Other Data (Model: FR-1)
Wingspan: 40 ft.
Length: 32 ft., 4 in.
Power Plant: One 1,350-hp Wright Cyclone; one 1,600-lb.-thrust General Electric turbojet
Loaded Weight: 11,652 lb.
Maximum Speed: 404 mph at 17,800 ft.

McDONNELL FH PHANTOM

McDonnell's FH Phantom made aviation history on July 21, 1946, when, in operations aboard the *Franklin D. Roosevelt*, it became the first pure jet to land on and take off from an aircraft carrier. It was also the Navy's first airplane to exceed a speed of 500 mph.

The story of the FH Phantom begins in 1943 when the Navy's Bureau of Aeronautics requested that the McDonnell Corporation begin development of an all-jet carrier-based fighter. Westinghouse Electric Corporation was called upon to design the two turbojet engines that would power the aircraft.

McDonnell engineers designed the Phantom as a low-wing monoplane with a tricycle landing gear. The wings were straight, that is, not swept back.

At the time that the plane's airframe was ready for flight testing, only one of the engines had been delivered, and it was installed on January 26, 1945. The first flight tests were conducted—with one engine still missing. So pleased were McDonnell officials with the results of the flights that advanced testing was conducted while the company continued to wait for the second engine.

After the aircraft's historic carrier takeoff and landing in 1946 (with two engines), the FH Phantom was ordered into production. But, as was the case with other aircraft of the time, orders were slashed when World War II ended in August, 1945.

Nevertheless, some Navy squadrons were supplied with the new plane. One of these, in simulated combat maneuvers aboard the carrier *Saipan* in May, 1948, made over 100 successful landings and takeoffs. When these trials had ended, there could be no doubt that a new era in naval aviation was at hand.

Other Data (Model: FH-1)
Wingspan: 40 ft., 9 in.
Length: 38 ft., 9 in.
Power Plant: Two 1,600-lb.-thrust Westinghouse turbojets
Loaded Weight: 12,035 lb.
Maximum Speed: 505 mph at 30,000 ft.

The FH Phantom was the Navy's first pure jet to land and take off from an aircraft carrier.

A Skyknight in flight, its wheels and arresting hooks down.

DOUGLAS F3D SKYKNIGHT

The Navy's first jet designed chiefly for night fighting, the F3D earned an outstanding combat record during the Korean War. According to official records, Skyknights destroyed more enemy aircraft over Korea than any other plane flown by Navy or Marine pilots.

A two-engine plane with a plump fuselage, the Skyknight made its first flight on March 23, 1948. The aircraft featured side-by-side seating for the two-man crew, which consisted of a pilot and radar operator.

The large amount of radar equipment the plane required for its role as a night-time interceptor was housed in the plane's bulbous nose. Its four 22-mm cannons were carried within the fuselage beneath the radar dome. This equipment, plus its two powerful Westinghouse turbojets, helped to make the F3D the biggest and heaviest aircraft ever designed for carrier-based operations up to that time.

This was a period before the development of a reliable ejection seat. Design engineers, however, provided F3D crew members with a method of escape in case of emergency. It consisted of a tunnel that led from the cockpit to a hatch beneath the fuselage. It enabled the pilot or radar man to drop from the underside of the plane. Although it sounds primitive, the system worked well, even at high speeds.

The first deliveries of the Skyknight to Navy and Marine squadrons began late in 1950, the same year that the Korean War broke out. The plane's first combat success was recorded on November 2, 1952, when a Skyknight brought down a Russian-built MiG-15. This marked the first time one jet plane had destroyed another during night operations.

One problem with the Skyknight was that it was unable to climb fast. To overcome this, a swept-wing version of the plane was designed. But this project was cancelled before the first plane was built.

The Navy ordered a total of 268 Skyknights. By the early 1960s, most of these had been replaced by more sophisticated aircraft. A few, however, continued to survive as missile carriers and radar trainers.

Other Data (Model: F3D-2)
Wingspan: 50 ft.
Length: 45 ft., 6 in.
Power Plant: Two Westinghouse turbojets delivering 6,800-lbs. total thrust
Loaded Weight: 26,850 lb.
Maximum Speed: 600 mph at 20,000 ft.

In the skies over Korea, no other Navy aircraft was as successful as the F3D Skyknight.

Panther's streamlined nose housed four 20-mm cannons.

GRUMMAN F9F PANTHER

During the closing days of World War II, the Navy asked the Grumman Company to develop a jet night-fighter. The aircraft that resulted never saw action in World War II, but it did carry the Navy's share of fighter operations during the Korean War.

In order to obtain the performance they wanted from their new fighter, Grumman engineers determined that it would require an engine with 6,000 pounds of thrust. The only jet engine available at that time was the Westinghouse J30, which had 1,500 pounds of thrust, one-quarter of what was needed.

Early designs for the F9F called for installing four engines in the plane's wings. Then Grumman learned of the Rolls-Royce Nene, an engine with 5,000 pounds of thrust. Grumman chose the Nene for the F9F's first power plant.

The first prototype for the F9F flew on August 16, 1948. A later prototype was equipped with an Allison engine, built by Pratt & Whitney as a replacement for the Nene. In still later models of the aircraft, other and more powerful engines were used. The first deliveries of production models of the F9F were made in May, 1949.

On July 3, 1950, Panthers from the carrier *Valley Forge* became the first Navy fighters to see action over Korea. Their assignment was to provide protection for a strike force made up of piston-powered AD Skyraiders and F4U Corsairs. On November 9, 1950, the F9F became the first carrier-based jet to shoot down another jet, a Russian-built MiG-15.

The first F9Fs had straight wings, but later models were of sweptwing design. The sweptwing version was known as the F9F Cougar. The Cougar replaced the Panther in many carrier squadrons and was chosen for use by the Blue Angels, the Navy's famed aerial acrobatic team, during the years from 1955 through 1958.

Other Data (Model: F9F-5)
Wingspan: 38 ft.
Length: 38 ft., 10 in.
Power Plant: One 6,520-lb.-thrust Pratt & Whitney turbojet
Loaded Weight: 18,721 lb.
Maximum Speed: 579 mph at 5,000 ft.

The Panther went through many modifications and conversions. This is the F9F-9.

LTV F-8 CRUSADER

In 1956, the Chance Vought aircraft company (which later became Ling-Temco-Vought, or LTV) won the coveted Collier Trophy for "the greatest achievement in aviation" during the preceding year. The basis for the award was the F-8 Crusader, the first carrier-based aircraft capable of speeds in excess of 1,000 mph.

The F-8's reputation as a speedster was established on March 25, 1955, the date of its first test flight, when the plane exceeded the speed of sound without a murmur. (The speed of sound can vary from 600 mph to 790 mph, depending on altitude and temperature.)

To enable the plane to make carrier landings and takeoffs, where less than 300 feet of runway was available, the F-8 was equipped with a hinged

Crusaders fly in tight formation over the carrier *Forrestal*.

Flight deck crew readies a Crusader for launch.

wing. On landings and takeoffs, when greater lift was required, the wing was lifted forward, which served to brake the aircraft. The wing was restored to its normal position for high-speed flight.

The Crusader bristled with armament. Besides its four forward-firing 20-mm cannons, the plane could be equipped with either four Sidewinder air-to-air missiles or up to 5,000 pounds of bombs.

Its big weapons package helped to make the F-8 extremely effective in the skies over Vietnam. Crusader pilots were credited with destroying 14 MiG-17s and 4 MiG-21s in air-to-air combat. In one engagement, a MiG-17 pilot ejected before a shot was fired, dramatic testimony to the F-8's aerial superiority.

Later models of the F-8 were called "push button" interceptors because of their advanced flight control systems. By using a "Mach hold" button, the pilot could maintain a specific climb angle at a constant rate of speed, something like the cruise control feature in the modern-day automobile.

Production of the Crusader ended with the F-8E, which first flew on June 30, 1961. All models of the plane were phased out during the mid-1960s.

Other Data (Model: F-8E)
Wingspan: 35 ft., 2 in.
Length: 54 ft., 6 in.
Power Plant: One 10,700-lb.-thrust Pratt & Whitney turbojet
Loaded Weight: 34,000 lb.
Maximum Speed: 1,120 mph at 40,000 ft.

GRUMMAN F11F TIGER

Another in the family of "cat" aircraft produced by the Grumman Company for the U.S. Navy, and which included the Hellcat, Bearcat, Cougar, and several others, the F11F Tiger was an efficient fighting plane that handled well. But it lacked the standout performance qualities of some of the Grumman fighters that had preceded it, and thus failed to enjoy a very long production life.

The design of the F11F was based upon that of the F9F Cougar, the sweptwing version of the F9F Panther (page 50). The aircraft was first flown on July 30, 1954.

Early tests demonstrated that the F11F Tiger needed more power. An afterburner (to burn the engine's exhaust gases) was added to provide greater thrust. Even with the afterburner, the F11F did not meet expectations. More changes and more testing followed.

It was March, 1957, before the F11F began to be assigned to fleet squadrons. Within two years, the Tiger was being phased out in favor of more advanced aircraft.

The F11F is probably best known for its many

While it did not build a brilliant record with the fleet, the F11F thrilled millions when it saw service with the Navy's Blue Angels.

years of service with the Blue Angels, the Navy's renowned flight demonstration team. It was chosen by the Blue Angels in 1957. With its excellent handling characteristics at supersonic speeds, the F11F was well suited for aerial acrobatics. The plane thrilled millions as the Blue Angels flashed through the skies.

Operational F11Fs were replaced with more advanced aircraft beginning in 1959. But the F11F continued to fly with the Blue Angels until 1968, the year it finally gave way to the McDonnell-Douglas F-4 Phantom II (page 56).

Other Data (Model: F11F-1)
Wingspan: 31 ft., 7½ in.
Length: 46 ft., 11¼ in.
Power Plant: One 7,450-lb.-thrust Wright turbojet
Loaded Weight: 22,160 lbs.
Maximum Speed: 750 mph at sea level

Noses painted with sharks'-teeth emblems, these F11Fs are reminiscent of World War II P-40s.

An F-4 Phantom II poised for launching aboard the carrier *Nimitz*.

McDONNELL-DOUGLAS F-4 PHANTOM II

The F-4 Phantom II, a twin-engine, two-man fighter and tactical strike aircraft, has been hailed as one of the finest air weapons of all time.

It is also one of the longest lasting. The F-4 was first flown on May 21, 1958, by Robert C. Little, chief test pilot for McDonnell Aircraft (the company's name before it merged with Douglas Aircraft). Its performance was so exceptional that the plane was ordered into full-scale production.

More than twenty years later, in 1979, when production ended, a total of 5,197 F-4s had been built. Less than one-quarter of that number—1,264 —were delivered to the Navy. The F-4 was also flown by the U.S. Air Force and Marine Corps. It was, in fact, the only fighter to serve simultaneously with all three services.

The F-4 was also the first aircraft to be flown by both of the nation's famous aerial demonstration teams, the Navy's Blue Angels and the Air Force's Thunderbirds.

The F-4 won wide acclaim overseas. It was operated by the air arms of no less than twelve nations.

The F-4 was originally developed in an effort to

provide the Navy with a high-performance fighter that was capable of operating day or night in any type of weather.

A two-man crew was decided upon to increase the plane's ability to detect enemy aircraft and attack at supersonic speeds. The pilot manned the controls from the front seat. The radar intercept officer was posted in the rear seat. He commanded the F-4's sophisticated radar system and also aided in the delivery of the many different weapons the plane carried.

Two engines were chosen for the plane instead of one, to insure greater safety for the crew.

From the first, the Phantom was a record-breaker. It established a speed record of 1,390.24 mph over a course measuring 100 kilometers (about 62 miles). An F-4 raced from Los Angeles to New York, a distance of 2,445 miles, in 48 minutes, another record.

On November 22, 1961, the Phantom advanced the world's official speed record to 1,665.89 mph. And to show how fast the plane could climb, an F-4 zoomed to 98,425 feet in 371.4 seconds.

Although unsurpassed as a speedster, the Phantom could also be operated safely at a speed as low as 125 mph. This made it an excellent plane for carrier operations.

The F-4B was the first production model of the Phantom II to be delivered to Navy squadrons in quantity. These aircraft were equipped with auto-

A flight deck tractor moves among F-4 Phantom IIs aboard the carrier *Forrestal*.

matic systems which located the target, tracked the prey, and set up the attack.

The F-4B saw frequent and wide-ranging duty during the war in Vietnam. Not only did the plane serve the Navy as an all-weather fighter, it was used by the Marines as a fighter-bomber and interceptor.

The F-4J was the second major production Phantom ordered by the Navy. This version of the plane was first flown in 1966.

The F-4J boasted improved radar. The radar

Its mission completed, an F-4 Phantom II approaches the flight deck of the carrier *Nimitz*.

antenna was housed in a larger nose dome than that of the F-4B.

The F-4J could be armed with either Sidewinder heat-seeking missiles or Sparrow III radar-guided missiles. The four Sparrow III missiles the plane carried were "semi-submerged," that is, part of each missile fitted into a depression in the plane's fuselage. Once launched, the Sparrow III streaked toward its target at speeds faster than 1,500 mph.

In all, the F-4J was capable of carrying a total of 16,000 pounds of explosives. The Air Force's huge four-engine B-47 Stratojet bomber carried a weapons package of up to 20,000 pounds.

The Phantom served not only as a fighter and interceptor but also as a bomber and reconnaissance aircraft. It came to be known in every part of the world. It was frontline equipment with the air forces of Israel, England, Iran, Germany, and Japan.

Thanks to modernization programs that added years to the combat life of the F-4, the aircraft continued to fly with the U.S. Navy through the 1970s and into the 1980s.

By the beginning of the decade, the F-4 was being replaced in some squadrons by the Grumman F-14 Tomcat (page 62). Nevertheless, there were more than 2,700 F-4s in service around the world in 1985, and it was said that by the end of the century an estimated 2,000 Phantoms would still be flying.

Other Data (Model: F-4J)
Wingspan: 38 ft., 4¾ in.
Length: 58 ft., 3¾ in.
Power Plant: Two General Electric turbojets
 delivering 35,800-lbs. total thrust
Loaded Weight: 59,000 lb.
Maximum Speed: 1,584 (Mach 2.4) at 30,000 ft.

(Right) Aboard the carrier *Enterprise*, a Phantom II pilot and copilot man their aircraft for their first combat mission over Vietnam.

(Below) A total of 5,197 F-4 Phantom IIs were built. This is the 5,000th.

An F-5E Tiger II, wearing camouflage paint, in flight off the California coast.

NORTHROP F-5E TIGER II

Light in weight and highly maneuverable, Northrop's F-5E Tiger II has been assigned an odd job. In mock aerial combat at the Air Combat Maneuvering Range in Yuma, Arizona, the F-5E always plays the role of the "enemy" fighter. It is painted in the colors of the "enemy" nation. Its pilot duplicates the flight characteristics of the "enemy" plane.

The F-5E has won great respect as an opponent. Indeed, it is not unusual for the F-5E to emerge as the "winner," providing the "defeated" Navy pilot a valuable lesson.

The F-5E did not begin life as a performer playing bad-guy parts. The aircraft grew out of the Navy's experience in the Korean War during the early 1950s. This revealed the need for a low-cost, lightweight, high-performance fighter. Design work on the plane began in 1955.

Although the Navy lost interest in the project, Northrop continued its development work. Early models of the plane, designated the F-5A, were supplied to a good number of foreign countries.

These included Thailand, South Korea, the Philippines, Pakistan, Nationalist China, Ethiopia, Iran, Morocco, Libya, Greece, and South Vietnam. During this period, the plane was called the Freedom Fighter.

In 1972, an improved version of the aircraft—the F-5E Tiger II—became available. This aircraft also proved extremely popular with foreign air forces. F-5E Tiger IIs were ordered by South Korea, Thailand, Iran, Malaysia, Switzerland, Taiwan, Brazil, Chile, Saudi Arabia, Jordan, and South Vietnam.

When South Vietnam surrendered in 1975, a number of Tiger IIs were abandoned to the enemy. It is believed that these aircraft are still being operated by the Vietnamese.

The U.S. Navy purchased its first F-5E in 1974. The planes were assigned to the Naval Fighter Weapons School at the Naval Air Station, Miramar, California. There they were used as the "aggressors" in aerial combat duels involving the F-4 Phantom II (page 56) and the F-14 Tomcat (page 62). Navy pilots who have opposed the F-5E have described the plane as a "tough nut to crack."

Other Data (Model: F-5E)
Wingspan: 25 ft., 3 in.
Length: 47 ft., 2 in.
Power Plant: Two General Electric turbojets delivering 10,000-lbs. total thrust
Loaded Weight: 24,083 lbs.
Maximum Speed: 1,050 mph (Mach 1.6) at 36,000 ft.

Light in weight and very maneuverable, F-5Es often play the role of the enemy plane in mock aerial combat duels. Scoreboard at left registers victories.

Underside view of F-14. Plane is armed with six radar-guided air-to-air-rockets, plus a 20-mm cannon.

GRUMMAN F-14 TOMCAT

The F-14 Tomcat has been called the ultimate air-combat weapons system. To put it more simply, it is a superplane.

The F-14 dates to 1969, the year the Navy asked both the Grumman Company and McDonnell Douglas to design an aircraft that would be a successor to the F-4 Phantom II. Grumman won the competition with its proposal for a two-seat, two-engine aircraft with a variable-swept wing.

The wing angle of the Tomcat is positioned by means of an on-board computer. The position varies with the airplane's speed. The wings are fully extended during landings and takeoffs, when maximum lift is needed. They're swept back against the fuselage when the pilot wants speed. When the wings are swept back as far as they can go, the Tomcat's nose can be pulled up vertically, allowing the plane to climb like a rocket. It seems to be supported by its jet thrust.

With its 65-foot wingspan, the Tomcat is one of the biggest fighters in the world. It is also one of the most heavily armed. One of its basic weapons is the large and expensive radar-guided Phoenix missile. The plane is also equipped with the Sparrow and Sidewinder missiles, as well as an M-61 20-mm rotary cannon. The cannon is capable of firing 4,000 to 6,000 rounds a minute.

Like the Phantom II, the aircraft it is meant to

replace, the Tomcat has a radar operator aboard to control its complex weapons systems. He can track as many as six of the plane's Phoenix missiles at one time, guiding them to separate targets up to 80 miles away. He can, at the same time, keep track of 18 other targets.

The first Tomcats were delivered to the Navy in 1972. There were problems with the plane at first but they were gradually overcome. By the early 1980s, Tomcats were operating from nine aircraft carriers as well as naval air stations at Miramar, California, and Oceana, Virginia. In early 1985, the Navy announced a program to modernize the F-14, thus assuring the plane's continuing role as the Navy's frontline fighter well into the 1990s.

Other Data (Model: F-14A)
Wingspan: 38 ft., 2.4 in. (swept)
64 ft., 1.5 in. (unswept)
Length: 62 ft., 8 in.
Power Plant: Two Pratt & Whitney turbofans delivering 41,800-lbs. total thrust
Loaded Weight: 70,426 lbs.
Maximum Speed: 1,564 mph (Mach 2.34)

As one F-14, its wings extended, prepares for launching, a second Tomcat, its wings still folded, noses into position.

NAVY AIRCRAFT ON DISPLAY

Many of the famous aircraft described in this book (and in a companion volume, *Famous Navy Attack Planes*) are on display at the U.S. Naval Aviation Museum in Pensacola, Florida.

These include such aircraft as the Naval Aircraft Factory TS-1, the first Navy plane actually planned as a fighter; Grumman's F6F Hellcat, one of the most noted planes of World War II; and McDonnell Douglas' F-4 Phantom II and other modern jets.

The museum's display of more than forty aircraft also includes the NC-4 flying boat, the first airplane to fly the Atlantic Ocean. The feat was accomplished in 1919.

Besides actual aircraft, the museum exhibits more than two hundred models of aircraft and historic ships, plus paintings, aircraft engines, and historic artifacts.

The museum is also the home for the Naval Aviation Hall of Honor, which pays tribute to individuals who have made significant contributions to the history and development of naval aviation.

For more information, write: Director, U.S. Naval Aviation Museum, Naval Air Station, Pensacola, FL 32508.

While the Naval Aviation Museum in Pensacola is the only museum in the world devoted exclusively to naval aviation, famous Navy aircraft can also be seen, studied, and photographed at several other museums. These include:

- *Intrepid* Sea-Air-Space Museum (One Intrepid Square, West 46th Street and 12th Avenue, New York, NY 10036).
- National Air and Space Museum (Smithsonian Institution, Washington, DC 20560).
- New England Air Museum (Bradley International Airport, Windsor Locks, CT 06096).
- San Diego Aerospace Museum (2001 Pan American Plaza, Balboa Park, San Diego, CA 92101).
- U.S. Marine Corps Aviation Museum (Brown Field, Quantico, VA 22134).

Write to the museums for more information, including viewing hours and possible admission fees.